THIS RECYCLING JOURNAL
BELONGS TO

_____

_____

Dedication

This Recycling Notebook is dedicated to all the people out there who want to do their part by recycling and document their findings in the process.

You are my inspiration for producing books and I'm honored to be a part of keeping all of your Recycling notes and records organized.

This journal notebook will help you record your details about tracking your recycling projects.

Thoughtfully put together with these sections to record:
Date, How Did I Help The Environment Today, What Can I Reuse, My Recycling List, Today I Love Earth Because, & How. Did I Reduce My Waste?

How to Use this Book

The purpose of this book is to keep all of your Recycling notes all in one place. It will help keep you organized.

This Recycling Notebook Journal will allow you to accurately document every detail about your recycling projects. It's a great way to chart your course through doing your part to protect our Earth.

Here are examples of the prompts for you to fill in and write about your experience in this book:

1. Date

2. How Did I Help The Environment Today?

3. What Can I Reuse?

4. My Recycling List For Today

5. Today I Love Earth Because. . .

6. How Did I Reduce My Waste Today?

Recycle

'Love Earth'

DATE: _____

How did I Help The Environment Today?

☐

☐

☐

☐

☐

☐

Today I Love Earth Because...

_____
_____
_____

What Can I Re-use?

_____
_____
_____
_____
_____
_____
_____
_____

My Recycling List for Today

_____
_____
_____
_____
_____
_____
_____
_____
_____
_____
_____
_____

How Did I Reduce my Waste Today?

_____
_____
_____
_____
_____

Recycle

'Love Earth'

DATE: _____

How did I Help The Environment Today?

- [ ]
- [ ]
- [ ]
- [ ]
- [ ]
- [ ]

Today I Love Earth Because...

_____
_____
_____

What Can I Re-use?

_____
_____
_____
_____
_____
_____
_____
_____
_____

My Recycling List for Today

_____
_____
_____
_____
_____
_____
_____
_____
_____
_____

How Did I Reduce my Waste Today?

_____
_____
_____
_____

Recycle

'Love Earth'

DATE: _____

How did I Help The Environment Today?

☐

☐

☐

☐

☐

☐

Today I Love Earth Because...

What Can I Re-use?

My Recycling List for Today

How Did I Reduce my Waste Today?

Recycle 'Love Earth'

DATE: _____

How did I Help The Environment Today?

- ☐
- ☐
- ☐
- ☐
- ☐
- ☐

Today I Love Earth Because...

_____
_____
_____

What Can I Re-use?

_____
_____
_____
_____
_____
_____
_____
_____

My Recycling List for Today

_____
_____
_____
_____
_____
_____
_____
_____
_____
_____

How Did I Reduce my Waste Today?

_____
_____
_____
_____
_____

Recycle

'Love Earth'

DATE: _____

How did I Help The Environment Today?

☐

☐

☐

☐

☐

☐

Today I Love Earth Because...

_____
_____
_____

What Can I Re-use?

_____
_____
_____
_____
_____
_____
_____
_____

My Recycling List for Today

_____
_____
_____
_____
_____
_____
_____
_____
_____
_____

How Did I Reduce my Waste Today?

_____
_____
_____
_____

Recycle

'Love Earth'

DATE: _____

How did I Help The Environment Today?

- ☐
- ☐
- ☐
- ☐
- ☐
- ☐

What Can I Re-use?

_____
_____
_____
_____
_____
_____
_____
_____

My Recycling List for Today

_____
_____
_____
_____
_____
_____
_____
_____
_____
_____

Today I Love Earth Because...

_____
_____
_____

How Did I Reduce my Waste Today?

_____
_____
_____
_____

Recycle
'Love Earth'

DATE: _____

How did I Help The Environment Today?

- [ ]
- [ ]
- [ ]
- [ ]
- [ ]
- [ ]

Today I Love Earth Because...

_____
_____
_____

What Can I Re-use?

_____
_____
_____
_____
_____
_____
_____
_____
_____

My Recycling List for Today

_____
_____
_____
_____
_____
_____
_____
_____
_____
_____
_____

How Did I Reduce my Waste Today?

_____
_____
_____
_____

Recycle

'Love Earth'

DATE: _____

How did I Help The Environment Today?

☐

☐

☐

☐

☐

☐

Today I Love Earth Because...

_____
_____
_____

What Can I Re-use?

_____
_____
_____
_____
_____
_____
_____
_____
_____
_____

My Recycling List for Today

_____
_____
_____
_____
_____
_____
_____
_____
_____
_____
_____
_____

How Did I Reduce my Waste Today?

_____
_____
_____
_____

Recycle

'Love Earth'

DATE: _____

How did I Help The Environment Today?

- [ ]
- [ ]
- [ ]
- [ ]
- [ ]
- [ ]

Today I Love Earth Because...

_____

_____

_____

What Can I Re-use?

_____
_____
_____
_____
_____
_____
_____
_____
_____

My Recycling List for Today

_____
_____
_____
_____
_____
_____
_____
_____
_____
_____
_____
_____

How Did I Reduce my Waste Today?

_____
_____
_____
_____
_____

Recycle

'Love Earth'

DATE: _____

How did I Help The Environment Today?

- [ ]
- [ ]
- [ ]
- [ ]
- [ ]
- [ ]

What Can I Re-use?

_____
_____
_____
_____
_____
_____
_____
_____
_____

My Recycling List for Today

_____
_____
_____
_____
_____
_____
_____
_____
_____
_____
_____
_____

Today I Love Earth Because...

_____
_____
_____

How Did I Reduce my Waste Today?

_____
_____
_____
_____
_____

Recycle

'Love Earth'

DATE: _____

How did I Help The Environment Today?

- [ ]
- [ ]
- [ ]
- [ ]
- [ ]
- [ ]

What Can I Re-use?

_____
_____
_____
_____
_____
_____
_____

My Recycling List for Today

_____
_____
_____
_____
_____
_____
_____
_____
_____
_____

Today I Love Earth Because...

_____
_____
_____

How Did I Reduce my Waste Today?

_____
_____
_____
_____
_____

Recycle

'Love Earth'

DATE: _____

How did I Help The Environment Today?

- ☐
- ☐
- ☐
- ☐
- ☐
- ☐

Today I Love Earth Because...

_____

_____

_____

What Can I Re-use?

_____

_____

_____

_____

_____

_____

_____

My Recycling List for Today

_____

_____

_____

_____

_____

_____

_____

_____

_____

How Did I Reduce my Waste Today?

_____

_____

_____

_____

_____

Recycle

'Love Earth'

DATE: _____

How did I Help The Environment Today?

☐

☐

☐

☐

☐

☐

Today I Love Earth Because...

_____

_____

_____

What Can I Re-use?

_____
_____
_____
_____
_____
_____
_____
_____
_____

My Recycling List for Today

_____
_____
_____
_____
_____
_____
_____
_____
_____
_____

How Did I Reduce my Waste Today?

_____
_____
_____
_____
_____

Recycle

'Love Earth'

DATE: _____

How did I Help The Environment Today?

What Can I Re-use?

_____
_____
_____
_____
_____
_____
_____
_____
_____

☐

☐

☐

My Recycling List for Today

_____
_____
_____
_____
_____
_____

☐

☐

_____
_____
_____
_____

☐

_____
_____

Today I Love Earth Because...

_____
_____
_____

_____
_____

How Did I Reduce my Waste Today?

_____
_____
_____
_____

Recycle

'Love Earth'

DATE: _____

How did I Help The Environment Today?

- [ ]
- [ ]
- [ ]
- [ ]
- [ ]
- [ ]

Today I Love Earth Because...

_____
_____
_____

What Can I Re-use?

_____
_____
_____
_____
_____
_____
_____
_____

My Recycling List for Today

_____
_____
_____
_____
_____
_____
_____
_____
_____
_____
_____
_____

How Did I Reduce my Waste Today?

_____
_____
_____
_____
_____

Recycle

'Love Earth'

DATE: _____

How did I Help The Environment Today?

- [ ]
- [ ]
- [ ]
- [ ]
- [ ]
- [ ]

Today I Love Earth Because...

_____
_____
_____

What Can I Re-use?

_____
_____
_____
_____
_____
_____
_____
_____

My Recycling List for Today

_____
_____
_____
_____
_____
_____
_____
_____
_____
_____
_____

How Did I Reduce my Waste Today?

_____
_____
_____
_____

Recycle

'Love Earth'

DATE: _____

How did I Help The Environment Today?

☐

☐

☐

☐

☐

☐

What Can I Re-use?

_____
_____
_____
_____
_____
_____
_____
_____

My Recycling List for Today

_____
_____
_____
_____
_____
_____
_____
_____
_____
_____

Today I Love Earth Because...

_____
_____
_____

How Did I Reduce my Waste Today?

_____
_____
_____
_____
_____

Recycle

'Love Earth'

DATE: _____

How did I Help The Environment Today?

☐

☐

☐

☐

☐

☐

What Can I Re-use?

_____
_____
_____
_____
_____
_____
_____
_____

My Recycling List for Today

_____
_____
_____
_____
_____
_____
_____
_____
_____
_____

Today I Love Earth Because...

_____
_____
_____

How Did I Reduce my Waste Today?

_____
_____
_____
_____

Recycle 'Love Earth'

DATE: _____

How did I Help The Environment Today?

☐

☐

☐

☐

☐

☐

What Can I Re-use?

_____
_____
_____
_____
_____
_____
_____
_____
_____

My Recycling List for Today

_____
_____
_____
_____
_____
_____
_____
_____
_____
_____
_____
_____
_____

Today I Love Earth Because...

_____
_____
_____

How Did I Reduce my Waste Today?

_____
_____
_____
_____
_____

Recycle

'Love Earth'

DATE: _____

How did I Help The Environment Today?

☐

☐

☐

☐

☐

☐

What Can I Re-use?

_____
_____
_____
_____
_____
_____
_____
_____

My Recycling List for Today

_____
_____
_____
_____
_____
_____
_____
_____
_____
_____
_____

Today I Love Earth Because...

_____
_____
_____

How Did I Reduce my Waste Today?

_____
_____
_____
_____

Recycle

'Love Earth'

DATE: _____

How did I Help The Environment Today?

- ☐
- ☐
- ☐
- ☐
- ☐
- ☐

Today I Love Earth Because...

_____
_____
_____

What Can I Re-use?

_____
_____
_____
_____
_____
_____
_____
_____
_____
_____

My Recycling List for Today

_____
_____
_____
_____
_____
_____
_____
_____
_____
_____

How Did I Reduce my Waste Today?

_____
_____
_____
_____
_____

Recycle

'Love Earth'

DATE: _____

How did I Help The Environment Today?

☐

☐

☐

☐

☐

☐

What Can I Re-use?

_____
_____
_____
_____
_____
_____
_____
_____

My Recycling List for Today

_____
_____
_____
_____
_____
_____
_____
_____
_____
_____
_____

Today I Love Earth Because...

_____
_____
_____

How Did I Reduce my Waste Today?

_____
_____
_____
_____

Recycle

'Love Earth'

DATE: _____

How did I Help The Environment Today?

☐

☐

☐

☐

☐

☐

Today I Love Earth Because...

_____
_____
_____

What Can I Re-use?

_____
_____
_____
_____
_____
_____
_____

My Recycling List for Today

_____
_____
_____
_____
_____
_____
_____
_____
_____
_____

How Did I Reduce my Waste Today?

_____
_____
_____
_____
_____

Recycle

'Love Earth'

DATE: _____

How did I Help The Environment Today?

☐

☐

☐

☐

☐

☐

Today I Love Earth Because...

_____

_____

_____

What Can I Re-use?

_____

_____

_____

_____

_____

_____

_____

My Recycling List for Today

_____

_____

_____

_____

_____

_____

_____

_____

_____

How Did I Reduce my Waste Today?

_____

_____

_____

_____

Recycle

'Love Earth'

DATE: _____

How did I Help The Environment Today?

- [ ]
- [ ]
- [ ]
- [ ]
- [ ]
- [ ]

Today I Love Earth Because...

_____
_____
_____

What Can I Re-use?

_____
_____
_____
_____
_____
_____
_____
_____

My Recycling List for Today

_____
_____
_____
_____
_____
_____
_____
_____

How Did I Reduce my Waste Today?

_____
_____
_____
_____
_____

Recycle 'Love Earth'

DATE: _____

How did I Help The Environment Today?

- ☐
- ☐
- ☐
- ☐
- ☐
- ☐

Today I Love Earth Because...

_____

_____

_____

What Can I Re-use?

_____
_____
_____
_____
_____
_____
_____
_____
_____
_____

My Recycling List for Today

_____
_____
_____
_____
_____
_____
_____
_____
_____
_____

How Did I Reduce my Waste Today?

_____
_____
_____
_____

Recycle

'Love Earth'

DATE: _____

How did I Help The Environment Today?

☐

☐

☐

☐

☐

☐

What Can I Re-use?

_____
_____
_____
_____
_____
_____
_____
_____

My Recycling List for Today

_____
_____
_____
_____
_____
_____
_____
_____
_____
_____

Today I Love Earth Because...

_____
_____
_____

How Did I Reduce my Waste Today?

_____
_____
_____
_____
_____

Recycle 'Love Earth'

DATE: _____

How did I Help The Environment Today?

- ☐
- ☐
- ☐
- ☐
- ☐
- ☐

Today I Love Earth Because...

_____
_____
_____

What Can I Re-use?

_____
_____
_____
_____
_____
_____
_____
_____
_____

My Recycling List for Today

_____
_____
_____
_____
_____
_____
_____
_____
_____
_____
_____
_____
_____

How Did I Reduce my Waste Today?

_____
_____
_____
_____
_____

Recycle

'Love Earth'

DATE: _____

How did I Help The Environment Today?

☐

☐

☐

☐

☐

☐

What Can I Re-use?

_____
_____
_____
_____
_____
_____
_____
_____
_____

My Recycling List for Today

_____
_____
_____
_____
_____
_____
_____
_____
_____
_____
_____
_____

Today I Love Earth Because...

_____
_____
_____

How Did I Reduce my Waste Today?

_____
_____
_____
_____

Recycle

'Love Earth'

DATE: _____

How did I Help The Environment Today?

- [ ]
- [ ]
- [ ]
- [ ]
- [ ]
- [ ]

What Can I Re-use?

_____
_____
_____
_____
_____
_____
_____
_____

My Recycling List for Today

_____
_____
_____
_____
_____
_____
_____
_____
_____
_____

Today I Love Earth Because...

_____
_____
_____

How Did I Reduce my Waste Today?

_____
_____
_____
_____

Recycle

'Love Earth'

D A T E: _____

How did I Help The Environment Today?

- [ ]
- [ ]
- [ ]
- [ ]
- [ ]
- [ ]

Today I Love Earth Because...

_____
_____
_____

What Can I Re-use?

_____
_____
_____
_____
_____
_____
_____
_____

My Recycling List for Today

_____
_____
_____
_____
_____
_____
_____
_____
_____
_____

How Did I Reduce my Waste Today?

_____
_____
_____
_____
_____

Recycle

'Love Earth'

DATE: _____

How did I Help The Environment Today?

- ☐
- ☐
- ☐
- ☐
- ☐
- ☐

Today I Love Earth Because...

_____

_____

_____

What Can I Re-use?

_____
_____
_____
_____
_____
_____
_____
_____

My Recycling List for Today

_____
_____
_____
_____
_____
_____
_____
_____
_____
_____
_____
_____

How Did I Reduce my Waste Today?

_____
_____
_____
_____

Recycle

'Love Earth'

D A T E : _____

How did I Help The Environment Today?

- [ ]
- [ ]
- [ ]
- [ ]
- [ ]
- [ ]

What Can I Re-use?

_____
_____
_____
_____
_____
_____
_____
_____
_____

My Recycling List for Today

_____
_____
_____
_____
_____
_____
_____
_____
_____
_____
_____

Today I Love Earth Because...

_____
_____
_____

How Did I Reduce my Waste Today?

_____
_____
_____
_____
_____

Recycle

'Love Earth'

DATE: _____

How did I Help The Environment Today?

☐

☐

☐

☐

☐

☐

Today I Love Earth Because...

_____

_____

_____

What Can I Re-use?

_____
_____
_____
_____
_____
_____
_____
_____
_____

My Recycling List for Today

_____
_____
_____
_____
_____
_____
_____
_____
_____
_____
_____
_____

How Did I Reduce my Waste Today?

_____
_____
_____
_____

Recycle

'Love Earth'

DATE: _____

How did I Help The Environment Today?

☐

☐

☐

☐

☐

☐

Today I Love Earth Because...

_____

_____

_____

What Can I Re-use?

_____
_____
_____
_____
_____
_____
_____

My Recycling List for Today

_____
_____
_____
_____
_____
_____
_____
_____
_____
_____

How Did I Reduce my Waste Today?

_____
_____
_____
_____

Recycle

'Love Earth'

DATE: _____

How did I Help The Environment Today?

- ☐
- ☐
- ☐
- ☐
- ☐
- ☐

Today I Love Earth Because...

_____
_____
_____

What Can I Re-use?

_____
_____
_____
_____
_____
_____
_____

My Recycling List for Today

_____
_____
_____
_____
_____
_____
_____
_____
_____
_____

How Did I Reduce my Waste Today?

_____
_____
_____
_____

Recycle

'Love Earth'

DATE: _____

How did I Help The Environment Today?

☐

☐

☐

☐

☐

☐

Today I Love Earth Because...

_____

_____

_____

What Can I Re-use?

_____
_____
_____
_____
_____
_____
_____
_____
_____
_____

My Recycling List for Today

_____
_____
_____
_____
_____
_____
_____
_____
_____
_____
_____
_____

How Did I Reduce my Waste Today?

_____
_____
_____
_____
_____

Recycle

'Love Earth'

DATE: _____

How did I Help The Environment Today?

☐

☐

☐

☐

☐

☐

What Can I Re-use?

_____
_____
_____
_____
_____
_____
_____
_____
_____

My Recycling List for Today

_____
_____
_____
_____
_____
_____
_____
_____
_____
_____
_____
_____

Today I Love Earth Because...

_____
_____
_____

How Did I Reduce my Waste Today?

_____
_____
_____
_____

Recycle

'Love Earth'

DATE: _____

How did I Help The Environment Today?

- ☐
- ☐
- ☐
- ☐
- ☐
- ☐

What Can I Re-use?

_____
_____
_____
_____
_____
_____
_____
_____

My Recycling List for Today

_____
_____
_____
_____
_____
_____
_____
_____
_____
_____
_____

Today I Love Earth Because...

_____
_____
_____

How Did I Reduce my Waste Today?

_____
_____
_____
_____

Recycle

'Love Earth'

DATE: _____

How did I Help The Environment Today?

☐

☐

☐

☐

☐

☐

What Can I Re-use?

_____
_____
_____
_____
_____
_____
_____
_____
_____
_____
_____

My Recycling List for Today

_____
_____
_____
_____
_____
_____
_____
_____
_____
_____
_____
_____
_____

Today I Love Earth Because...

_____
_____
_____

How Did I Reduce my Waste Today?

_____
_____
_____
_____
_____

Recycle

'Love Earth'

DATE: _____

How did I Help The Environment Today?

☐

☐

☐

☐

☐

☐

Today I Love Earth Because...

_____
_____
_____

What Can I Re-use?

_____
_____
_____
_____
_____
_____
_____

My Recycling List for Today

_____
_____
_____
_____
_____
_____
_____
_____
_____
_____

How Did I Reduce my Waste Today?

_____
_____
_____
_____

Recycle

'Love Earth'

DATE: _____

How did I Help The Environment Today?

- ☐
- ☐
- ☐
- ☐
- ☐
- ☐

Today I Love Earth Because...

_____

_____

_____

What Can I Re-use?

_____

_____

_____

_____

_____

_____

_____

My Recycling List for Today

_____

_____

_____

_____

_____

_____

_____

_____

_____

How Did I Reduce my Waste Today?

_____

_____

_____

_____

Recycle

'Love Earth'

DATE: _____

How did I Help The Environment Today?

☐

☐

☐

☐

☐

☐

Today I Love Earth Because...

_____

_____

_____

What Can I Re-use?

_____
_____
_____
_____
_____
_____
_____
_____

My Recycling List for Today

_____
_____
_____
_____
_____
_____
_____
_____
_____
_____
_____

How Did I Reduce my Waste Today?

_____
_____
_____
_____

Recycle

'Love Earth'

DATE: _____

How did I Help The Environment Today?

- [ ]
- [ ]
- [ ]
- [ ]
- [ ]
- [ ]

What Can I Re-use?

_____
_____
_____
_____
_____
_____
_____
_____
_____
_____

My Recycling List for Today

_____
_____
_____
_____
_____
_____
_____
_____
_____
_____
_____
_____

Today I Love Earth Because...

_____
_____
_____

How Did I Reduce my Waste Today?

_____
_____
_____
_____

Recycle

'Love Earth'

DATE: _____

How did I Help The Environment Today?

☐

☐

☐

☐

☐

☐

Today I Love Earth Because...

_____
_____
_____

What Can I Re-use?

_____
_____
_____
_____
_____
_____
_____
_____
_____

My Recycling List for Today

_____
_____
_____
_____
_____
_____
_____
_____
_____
_____
_____

How Did I Reduce my Waste Today?

_____
_____
_____
_____
_____

Recycle

'Love Earth'

D A T E: _____

How did I Help The Environment Today?

☐

☐

☐

☐

☐

☐

Today I Love Earth Because...

_____

_____

_____

What Can I Re-use?

_____
_____
_____
_____
_____
_____
_____
_____

My Recycling List for Today

_____
_____
_____
_____
_____
_____
_____
_____
_____
_____
_____

How Did I Reduce my Waste Today?

_____
_____
_____
_____

Recycle

'Love Earth'

D A T E: _____

How did I Help The Environment Today?

☐

☐

☐

☐

☐

☐

Today I Love Earth Because...

_____
_____
_____

What Can I Re-use?

_____
_____
_____
_____
_____
_____
_____
_____

My Recycling List for Today

_____
_____
_____
_____
_____
_____
_____
_____
_____
_____

How Did I Reduce my Waste Today?

_____
_____
_____
_____
_____

Recycle

'Love Earth'

DATE: _____

How did I Help The Environment Today?

- [ ]
- [ ]
- [ ]
- [ ]
- [ ]
- [ ]

Today I Love Earth Because...

_____
_____
_____

What Can I Re-use?

_____
_____
_____
_____
_____
_____
_____
_____

My Recycling List for Today

_____
_____
_____
_____
_____
_____
_____
_____
_____
_____
_____

How Did I Reduce my Waste Today?

_____
_____
_____
_____

Recycle

'Love Earth'

DATE: _____

How did I Help The Environment Today?

☐

☐

☐

☐

☐

☐

What Can I Re-use?

_____
_____
_____
_____
_____
_____
_____
_____
_____

My Recycling List for Today

_____
_____
_____
_____
_____
_____
_____
_____
_____
_____
_____
_____

Today I Love Earth Because...

_____
_____
_____

How Did I Reduce my Waste Today?

_____
_____
_____
_____
_____

Recycle

'Love Earth'

DATE: _____

How did I Help The Environment Today?

- [ ]
- [ ]
- [ ]
- [ ]
- [ ]
- [ ]

What Can I Re-use?

_____
_____
_____
_____
_____
_____
_____
_____
_____

My Recycling List for Today

_____
_____
_____
_____
_____
_____
_____
_____
_____
_____
_____
_____
_____

Today I Love Earth Because...

_____
_____
_____

How Did I Reduce my Waste Today?

_____
_____
_____
_____

Recycle

'Love Earth'

DATE: _____

How did I Help The Environment Today?

- ☐
- ☐
- ☐
- ☐
- ☐
- ☐

What Can I Re-use?

_____
_____
_____
_____
_____
_____
_____
_____
_____

My Recycling List for Today

_____
_____
_____
_____
_____
_____
_____
_____
_____
_____
_____

Today I Love Earth Because...

_____
_____
_____

How Did I Reduce my Waste Today?

_____
_____
_____
_____
_____

Recycle

'Love Earth'

DATE: _____

How did I Help The Environment Today?

☐

☐

☐

☐

☐

☐

What Can I Re-use?

_____
_____
_____
_____
_____
_____
_____
_____
_____

My Recycling List for Today

_____
_____
_____
_____
_____
_____
_____
_____
_____
_____
_____
_____

Today I Love Earth Because...

_____
_____
_____

How Did I Reduce my Waste Today?

_____
_____
_____
_____
_____

Recycle

'Love Earth'

DATE: _____

How did I Help The Environment Today?

☐

☐

☐

☐

☐

☐

What Can I Re-use?

_____
_____
_____
_____
_____
_____
_____
_____

My Recycling List for Today

_____
_____
_____
_____
_____
_____
_____
_____
_____
_____
_____

Today I Love Earth Because...

_____
_____
_____

How Did I Reduce my Waste Today?

_____
_____
_____
_____
_____

Recycle

'Love Earth'

DATE: _____

How did I Help The Environment Today?

☐

☐

☐

☐

☐

☐

Today I Love Earth Because...

_____

_____

_____

What Can I Re-use?

_____
_____
_____
_____
_____
_____
_____
_____

My Recycling List for Today

_____
_____
_____
_____
_____
_____
_____
_____
_____
_____

How Did I Reduce my Waste Today?

_____
_____
_____
_____

Recycle

'Love Earth'

DATE: _____

How did I Help The Environment Today?

☐

☐

☐

☐

☐

☐

Today I Love Earth Because...

_____

_____

_____

What Can I Re-use?

_____
_____
_____
_____
_____
_____
_____
_____
_____

My Recycling List for Today

_____
_____
_____
_____
_____
_____
_____
_____
_____
_____
_____

How Did I Reduce my Waste Today?

_____
_____
_____
_____
_____

Recycle

'Love Earth'

DATE: _____

How did I Help The Environment Today?

What Can I Re-use?

☐

☐

☐

☐

☐

☐

My Recycling List for Today

Today I Love Earth Because...

How Did I Reduce my Waste Today?

Recycle

'Love Earth'

DATE: _____

How did I Help The Environment Today?

☐

☐

☐

☐

☐

☐

Today I Love Earth Because...

_____

_____

_____

What Can I Re-use?

_____
_____
_____
_____
_____
_____
_____

My Recycling List for Today

_____
_____
_____
_____
_____
_____
_____
_____
_____
_____
_____

How Did I Reduce my Waste Today?

_____
_____
_____
_____

Recycle

'Love Earth'

DATE: _____

How did I Help The Environment Today?

- ☐
- ☐
- ☐
- ☐
- ☐
- ☐

What Can I Re-use?

_____
_____
_____
_____
_____
_____
_____
_____
_____
_____

My Recycling List for Today

_____
_____
_____
_____
_____
_____
_____
_____
_____
_____
_____
_____

Today I Love Earth Because...

_____
_____
_____

How Did I Reduce my Waste Today?

_____
_____
_____
_____

Recycle

'Love Earth'

D A T E: _____

How did I Help The Environment Today?

☐

☐

☐

☐

☐

☐

Today I Love Earth Because...

_____
_____
_____

What Can I Re-use?

_____
_____
_____
_____
_____
_____
_____

My Recycling List for Today

_____
_____
_____
_____
_____
_____
_____
_____
_____

How Did I Reduce my Waste Today?

_____
_____
_____
_____

Recycle

'Love Earth'

DATE: _____

How did I Help The Environment Today?

- [ ]
- [ ]
- [ ]
- [ ]
- [ ]
- [ ]

Today I Love Earth Because...

_____
_____
_____

What Can I Re-use?

_____
_____
_____
_____
_____
_____
_____
_____

My Recycling List for Today

_____
_____
_____
_____
_____
_____
_____
_____
_____
_____

How Did I Reduce my Waste Today?

_____
_____
_____
_____
_____

Recycle

'Love Earth'

D A T E: _____

How did I Help The Environment Today?

☐

☐

☐

☐

☐

☐

What Can I Re-use?

_____
_____
_____
_____
_____
_____
_____
_____
_____

My Recycling List for Today

_____
_____
_____
_____
_____
_____
_____
_____
_____
_____

Today I Love Earth Because...

_____
_____
_____

How Did I Reduce my Waste Today?

_____
_____
_____
_____
_____

Recycle

'Love Earth'

DATE: _____

How did I Help The Environment Today?

- [ ]
- [ ]
- [ ]
- [ ]
- [ ]
- [ ]

Today I Love Earth Because...

_____

_____

_____

What Can I Re-use?

_____
_____
_____
_____
_____
_____
_____
_____

My Recycling List for Today

_____
_____
_____
_____
_____
_____
_____
_____
_____
_____

How Did I Reduce my Waste Today?

_____
_____
_____
_____
_____

Recycle

'Love Earth'

DATE: _____

How did I Help The Environment Today?

☐

☐

☐

☐

☐

☐

Today I Love Earth Because...

_____

_____

_____

What Can I Re-use?

_____
_____
_____
_____
_____
_____
_____
_____

My Recycling List for Today

_____
_____
_____
_____
_____
_____
_____
_____
_____
_____

How Did I Reduce my Waste Today?

_____
_____
_____
_____

Recycle

'Love Earth'

DATE: _____

How did I Help The Environment Today?

☐

☐

☐

☐

☐

☐

What Can I Re-use?

_____
_____
_____
_____
_____
_____
_____
_____
_____

My Recycling List for Today

_____
_____
_____
_____
_____
_____
_____
_____
_____
_____
_____
_____

Today I Love Earth Because...

_____
_____
_____

How Did I Reduce my Waste Today?

_____
_____
_____
_____
_____

Recycle

'Love Earth'

DATE: _____

How did I Help The Environment Today?

- ☐
- ☐
- ☐
- ☐
- ☐
- ☐

What Can I Re-use?

_____
_____
_____
_____
_____
_____
_____
_____

My Recycling List for Today

_____
_____
_____
_____
_____
_____
_____
_____
_____
_____

Today I Love Earth Because...

_____
_____

How Did I Reduce my Waste Today?

_____
_____
_____
_____
_____

Recycle

'Love Earth'

D A T E: _____

How did I Help The Environment Today?

- [ ]
- [ ]
- [ ]
- [ ]
- [ ]
- [ ]

Today I Love Earth Because...

_____

_____

_____

What Can I Re-use?

_____

_____

_____

_____

_____

_____

_____

My Recycling List for Today

_____

_____

_____

_____

_____

_____

_____

_____

_____

_____

How Did I Reduce my Waste Today?

_____

_____

_____

_____

_____

Recycle

'Love Earth'

DATE: _____

How did I Help The Environment Today?

☐

☐

☐

☐

☐

☐

Today I Love Earth Because...

_____

_____

_____

What Can I Re-use?

_____
_____
_____
_____
_____
_____
_____
_____
_____

My Recycling List for Today

_____
_____
_____
_____
_____
_____
_____
_____
_____
_____
_____

How Did I Reduce my Waste Today?

_____
_____
_____
_____
_____

Recycle

'Love Earth'

DATE: _____

How did I Help The Environment Today?

☐

☐

☐

☐

☐

☐

What Can I Re-use?

_____
_____
_____
_____
_____
_____
_____
_____
_____

My Recycling List for Today

_____
_____
_____
_____
_____
_____
_____
_____
_____
_____
_____
_____

Today I Love Earth Because...

_____
_____
_____

How Did I Reduce my Waste Today?

_____
_____
_____
_____

Recycle

'Love Earth'

DATE: _____

How did I Help The Environment Today?

☐

☐

☐

☐

☐

☐

Today I Love Earth Because...

_____
_____
_____

What Can I Re-use?

_____
_____
_____
_____
_____
_____
_____
_____

My Recycling List for Today

_____
_____
_____
_____
_____
_____
_____
_____
_____

How Did I Reduce my Waste Today?

_____
_____
_____
_____
_____

Recycle

'Love Earth'

DATE: _____

How did I Help The Environment Today?

- [ ]
- [ ]
- [ ]
- [ ]
- [ ]
- [ ]

Today I Love Earth Because...

_____

_____

_____

What Can I Re-use?

_____
_____
_____
_____
_____
_____
_____
_____
_____
_____

My Recycling List for Today

_____
_____
_____
_____
_____
_____
_____
_____
_____
_____
_____
_____

How Did I Reduce my Waste Today?

_____
_____
_____
_____
_____

Recycle

'Love Earth'

DATE: _____

How did I Help The Environment Today?

- ☐
- ☐
- ☐
- ☐
- ☐
- ☐

Today I Love Earth Because...

_____
_____
_____

What Can I Re-use?

_____
_____
_____
_____
_____
_____
_____
_____

My Recycling List for Today

_____
_____
_____
_____
_____
_____
_____
_____
_____
_____
_____

How Did I Reduce my Waste Today?

_____
_____
_____
_____

Recycle

'Love Earth'

DATE: _____

How did I Help The Environment Today?

☐

☐

☐

☐

☐

☐

Today I Love Earth Because...

_____

_____

_____

What Can I Re-use?

_____
_____
_____
_____
_____
_____
_____
_____
_____

My Recycling List for Today

_____
_____
_____
_____
_____
_____
_____
_____
_____
_____
_____

How Did I Reduce my Waste Today?

_____
_____
_____
_____
_____

Recycle

'Love Earth'

DATE: _____

How did I Help The Environment Today?

☐

☐

☐

☐

☐

☐

Today I Love Earth Because...

_____

_____

_____

What Can I Re-use?

_____
_____
_____
_____
_____
_____
_____
_____

My Recycling List for Today

_____
_____
_____
_____
_____
_____
_____
_____
_____
_____

How Did I Reduce my Waste Today?

_____
_____
_____
_____
_____

Recycle

'Love Earth'

D A T E: _____

How did I Help The Environment Today?

- ☐
- ☐
- ☐
- ☐
- ☐
- ☐

Today I Love Earth Because...

_____
_____

What Can I Re-use?

_____
_____
_____
_____
_____
_____
_____
_____

My Recycling List for Today

_____
_____
_____
_____
_____
_____
_____
_____
_____
_____

How Did I Reduce my Waste Today?

_____
_____
_____
_____
_____

Recycle 'Love Earth'

DATE: _____

How did I Help The Environment Today?

☐

☐

☐

☐

☐

☐

Today I Love Earth Because...

_____
_____
_____

What Can I Re-use?

_____
_____
_____
_____
_____
_____
_____

My Recycling List for Today

_____
_____
_____
_____
_____
_____
_____
_____
_____
_____

How Did I Reduce my Waste Today?

_____
_____
_____
_____
_____

Recycle

'Love Earth'

DATE: _____

How did I Help The Environment Today?

- [ ]
- [ ]
- [ ]
- [ ]
- [ ]
- [ ]

What Can I Re-use?

_____
_____
_____
_____
_____
_____
_____
_____
_____

My Recycling List for Today

_____
_____
_____
_____
_____
_____
_____
_____
_____
_____
_____

Today I Love Earth Because...

_____
_____
_____

How Did I Reduce my Waste Today?

_____
_____
_____
_____

Recycle

'Love Earth'

DATE: _____

How did I Help The Environment Today?

- ☐
- ☐
- ☐
- ☐
- ☐
- ☐

What Can I Re-use?

_____
_____
_____
_____
_____
_____
_____
_____

My Recycling List for Today

_____
_____
_____
_____
_____
_____
_____
_____
_____
_____

Today I Love Earth Because...

_____
_____
_____

How Did I Reduce my Waste Today?

_____
_____
_____
_____
_____

Recycle

'Love Earth'

DATE: _____

How did I Help The Environment Today?

☐

☐

☐

☐

☐

☐

Today I Love Earth Because...

_____

_____

_____

What Can I Re-use?

_____
_____
_____
_____
_____
_____
_____
_____
_____

My Recycling List for Today

_____
_____
_____
_____
_____
_____
_____
_____
_____
_____

How Did I Reduce my Waste Today?

_____
_____
_____
_____

Recycle

'Love Earth'

DATE: _____

How did I Help The Environment Today?

☐

☐

☐

☐

☐

☐

Today I Love Earth Because...

_____

_____

_____

What Can I Re-use?

_____
_____
_____
_____
_____
_____
_____
_____
_____
_____

My Recycling List for Today

_____
_____
_____
_____
_____
_____
_____
_____
_____
_____
_____

How Did I Reduce my Waste Today?

_____
_____
_____
_____
_____

Recycle 'Love Earth'

DATE: _____

How did I Help The Environment Today?

☐

☐

☐

☐

☐

☐

Today I Love Earth Because...

_____

_____

_____

What Can I Re-use?

_____
_____
_____
_____
_____
_____
_____
_____

My Recycling List for Today

_____
_____
_____
_____
_____
_____
_____
_____
_____
_____
_____

How Did I Reduce my Waste Today?

_____
_____
_____
_____

Recycle

'Love Earth'

DATE: _____

How did I Help The Environment Today?

☐

☐

☐

☐

☐

☐

What Can I Re-use?

_____
_____
_____
_____
_____
_____
_____

My Recycling List for Today

_____
_____
_____
_____
_____
_____
_____
_____
_____
_____
_____
_____
_____

Today I Love Earth Because...

_____
_____
_____

How Did I Reduce my Waste Today?

_____
_____
_____
_____
_____

Recycle

'Love Earth'

DATE: _____

How did I Help The Environment Today?

☐

☐

☐

☐

☐

☐

Today I Love Earth Because...

_____

_____

_____

What Can I Re-use?

_____
_____
_____
_____
_____
_____
_____
_____
_____

My Recycling List for Today

_____
_____
_____
_____
_____
_____
_____
_____
_____
_____
_____
_____
_____

How Did I Reduce my Waste Today?

_____

_____

_____

_____

Recycle

'Love Earth'

DATE: _____

How did I Help The Environment Today?

☐

☐

☐

☐

☐

☐

What Can I Re-use?

_____
_____
_____
_____
_____
_____
_____
_____

My Recycling List for Today

_____
_____
_____
_____
_____
_____
_____
_____
_____
_____
_____
_____

Today I Love Earth Because...

_____
_____
_____

How Did I Reduce my Waste Today?

_____
_____
_____
_____
_____

Recycle

'Love Earth'

DATE: _____

How did I Help The Environment Today?

- [ ]
- [ ]
- [ ]
- [ ]
- [ ]
- [ ]

What Can I Re-use?

_____
_____
_____
_____
_____
_____
_____
_____
_____

My Recycling List for Today

_____
_____
_____
_____
_____
_____
_____
_____
_____
_____

Today I Love Earth Because...

_____
_____
_____

How Did I Reduce my Waste Today?

_____
_____
_____
_____

Recycle

'Love Earth'

DATE: _____

How did I Help The Environment Today?

☐

☐

☐

☐

☐

☐

Today I Love Earth Because...

_____
_____
_____

What Can I Re-use?

_____
_____
_____
_____
_____
_____
_____
_____

My Recycling List for Today

_____
_____
_____
_____
_____
_____
_____
_____
_____

How Did I Reduce my Waste Today?

_____
_____
_____
_____
_____

Recycle

'Love Earth'

D A T E: _____

How did I Help The Environment Today?

- [ ]
- [ ]
- [ ]
- [ ]
- [ ]
- [ ]

What Can I Re-use?

_____
_____
_____
_____
_____
_____
_____
_____
_____

My Recycling List for Today

_____
_____
_____
_____
_____
_____
_____
_____
_____
_____
_____

Today I Love Earth Because...

_____
_____
_____

How Did I Reduce my Waste Today?

_____
_____
_____
_____
_____

Recycle

'Love Earth'

DATE: _____

How did I Help The Environment Today?

☐

☐

☐

☐

☐

☐

Today I Love Earth Because...

_____

_____

_____

What Can I Re-use?

_____
_____
_____
_____
_____
_____
_____
_____

My Recycling List for Today

_____
_____
_____
_____
_____
_____
_____
_____
_____
_____

How Did I Reduce my Waste Today?

_____
_____
_____
_____

Recycle

'Love Earth'

DATE: _____

How did I Help The Environment Today?

What Can I Re-use?

_____
_____
_____
_____
_____
_____
_____
_____
_____

My Recycling List for Today

_____
_____
_____
_____
_____
_____
_____
_____
_____
_____
_____

☐

☐

☐

☐

☐

☐

Today I Love Earth Because...

_____
_____
_____

How Did I Reduce my Waste Today?

_____
_____
_____
_____
_____

Recycle

'Love Earth'

DATE: _____

How did I Help The Environment Today?

☐

☐

☐

☐

☐

☐

Today I Love Earth Because...

What Can I Re-use?

My Recycling List for Today

How Did I Reduce my Waste Today?

Recycle

'Love Earth'

DATE: _____

How did I Help The Environment Today?

☐

☐

☐

☐

☐

☐

What Can I Re-use?

_____
_____
_____
_____
_____
_____
_____
_____

My Recycling List for Today

_____
_____
_____
_____
_____
_____
_____
_____
_____

Today I Love Earth Because...

_____
_____
_____

How Did I Reduce my Waste Today?

_____
_____
_____
_____

Recycle

'Love Earth'

DATE: _____

How did I Help The Environment Today?

☐

☐

☐

☐

☐

☐

Today I Love Earth Because...

_____

_____

_____

What Can I Re-use?

_____
_____
_____
_____
_____
_____
_____
_____
_____
_____
_____

My Recycling List for Today

_____
_____
_____
_____
_____
_____
_____
_____
_____
_____

How Did I Reduce my Waste Today?

_____
_____
_____
_____
_____

Recycle

'Love Earth'

DATE: _____

How did I Help The Environment Today?

☐

☐

☐

☐

☐

☐

Today I Love Earth Because...

_____

_____

_____

What Can I Re-use?

_____
_____
_____
_____
_____
_____
_____
_____
_____

My Recycling List for Today

_____
_____
_____
_____
_____
_____
_____
_____
_____
_____

How Did I Reduce my Waste Today?

_____
_____
_____
_____
_____

Recycle

'Love Earth'

DATE: _____

How did I Help The Environment Today?

☐

☐

☐

☐

☐

☐

What Can I Re-use?

_____
_____
_____
_____
_____
_____
_____
_____

My Recycling List for Today

_____
_____
_____
_____
_____
_____
_____
_____
_____
_____

Today I Love Earth Because...

_____
_____
_____

How Did I Reduce my Waste Today?

_____
_____
_____
_____
_____

Recycle

'Love Earth'

DATE: _____

How did I Help The Environment Today?

☐

☐

☐

☐

☐

☐

Today I Love Earth Because...

_____
_____
_____

What Can I Re-use?

_____
_____
_____
_____
_____
_____
_____
_____

My Recycling List for Today

_____
_____
_____
_____
_____
_____
_____
_____
_____
_____
_____

How Did I Reduce my Waste Today?

_____
_____
_____
_____

Recycle

'Love Earth'

DATE: _____

How did I Help The Environment Today?

☐

☐

☐

☐

☐

☐

Today I Love Earth Because...

_____

_____

_____

What Can I Re-use?

_____
_____
_____
_____
_____
_____
_____

My Recycling List for Today

_____
_____
_____
_____
_____
_____
_____
_____
_____

How Did I Reduce my Waste Today?

_____
_____
_____
_____
_____

Recycle

'Love Earth'

DATE: _____

How did I Help The Environment Today?

☐

☐

☐

☐

☐

☐

Today I Love Earth Because...

_____

_____

_____

What Can I Re-use?

_____
_____
_____
_____
_____
_____
_____
_____

My Recycling List for Today

_____
_____
_____
_____
_____
_____
_____
_____
_____
_____
_____

How Did I Reduce my Waste Today?

_____
_____
_____
_____

Recycle

'Love Earth'

DATE: _____

How did I Help The Environment Today?

☐

☐

☐

☐

☐

☐

Today I Love Earth Because...

_____

_____

_____

What Can I Re-use?

_____
_____
_____
_____
_____
_____
_____
_____
_____

My Recycling List for Today

_____
_____
_____
_____
_____
_____
_____
_____
_____
_____
_____
_____

How Did I Reduce my Waste Today?

_____
_____
_____
_____
_____

Recycle

'Love Earth'

DATE: _____

How did I Help The Environment Today?

- ☐
- ☐
- ☐
- ☐
- ☐
- ☐

Today I Love Earth Because...

_____
_____
_____

What Can I Re-use?

_____
_____
_____
_____
_____
_____
_____
_____

My Recycling List for Today

_____
_____
_____
_____
_____
_____
_____
_____
_____
_____

How Did I Reduce my Waste Today?

_____
_____
_____
_____

Recycle

'Love Earth'

D A T E : _____

How did I Help The Environment Today?

☐

☐

☐

☐

☐

☐

Today I Love Earth Because...

_____

_____

_____

What Can I Re-use?

_____
_____
_____
_____
_____
_____
_____
_____

My Recycling List for Today

_____
_____
_____
_____
_____
_____
_____
_____
_____
_____
_____
_____

How Did I Reduce my Waste Today?

_____
_____
_____
_____

www.ingramcontent.com/pod-product-compliance
Lightning Source LLC
Chambersburg PA
CBHW060934220326
41597CB00020BA/3823